Available in MyStatLab™ for Your Introductory Statistics Courses

MyStatLab is the market-leading online resource for learning and teaching statistics.

Leverage the Power of StatCrunch

MyStatLab leverages the power of StatCrunch—powerful, web-based statistics software. Integrated into MyStatLab, students can easily analyze data from their exercises and etext. In addition, access to the full online community allows users to take advantage of a wide variety of resources and applications at www.statcrunch.com.

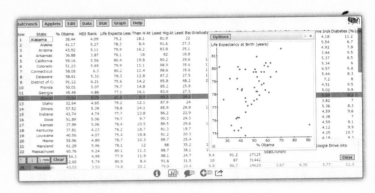

Bring Statistics to Life

Virtually flip coins, roll dice, draw cards, and interact with animations on your mobile device with the extensive menu of experiments and applets in StatCrunch. In addition to the author-created web apps that accompany the text, StatCrunch offers a number of ways to practice resampling procedures, such as permutation tests and bootstrap confidence intervals. StatCrunch is a complete and modern solution.

Real-World Statistics

MyStatLab video resources help foster conceptual understanding. StatTalk Videos, hosted by fun-loving statistician Andrew Vickers, demonstrate important statistical concepts through interesting stories and real-life events. This series of 24 videos includes assignable questions built in MyStatLab and an instructor's guide.

www.mystatlab.com

A Guide to Learning From the Art in This Text

We use color to help distinguish between the different shapes that graphs may take:

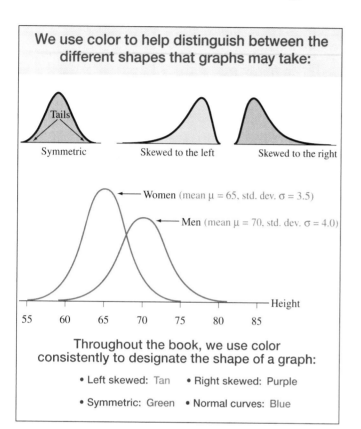

Throughout the book, we use color consistently to designate the shape of a graph:

- Left skewed: Tan
- Right skewed: Purple
- Symmetric: Green
- Normal curves: Blue

And between important measures such as the sample median and sample mean

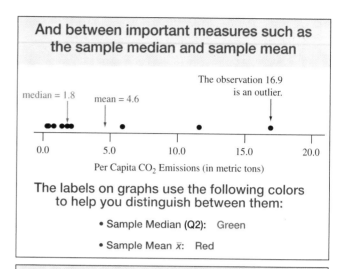

The labels on graphs use the following colors to help you distinguish between them:

- Sample Median (Q2): Green
- Sample Mean $\bar{x}$: Red

and between some of the most important statistics and parameters

	Sample Statistic	Population Parameter
Mean	$\bar{x}$	μ
Standard Deviation	s	σ
Proportion	$\hat{p}$	p

We show **Sampling Distributions of Sample Proportions** in blue because the normal distribution is used to describe the sampling distribution of $\hat{p}$.

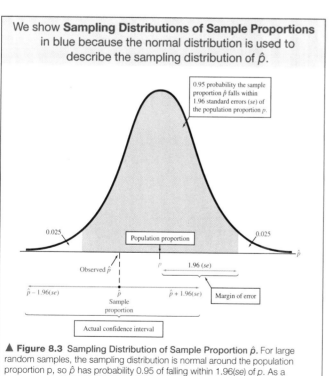

▲ **Figure 8.3** Sampling Distribution of Sample Proportion $\hat{p}$. For large random samples, the sampling distribution is normal around the population proportion p, so $\hat{p}$ has probability 0.95 of falling within 1.96(se) of p. As a consequence, $\hat{p} \pm 1.96(se)$ is a 95% confidence interval for p. **Question** Why is the confidence interval $\hat{p} \pm 1.96(se)$ instead of $p \pm 1.96(se)$?

We show **Sampling Distributions of Sample Means** in green because the symmetric t distribution is used to describe the sampling distribution of $\bar{x}$.

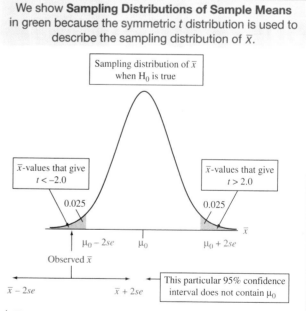

▲ **Figure 9.8** Relation between Confidence Interval and Significance Test. With large samples, if the sample mean falls more than about two standard errors from μ_0, then μ_0 does not fall in the 95% confidence interval and also μ_0 is rejected in a test at the 0.05 significance level. **Question** Inference about proportions does not have an *exact* equivalence between confidence intervals and tests. Why? (*Hint*: Are the same standard error values used in the two methods?)

Read and think about the questions that appear in selected figures. The answers are given at the beginning of each Chapter Review section.

Editorial Director: Chris Hoag
Editor in Chief: Deirdre Lynch
Acquisitions Editor: Suzanna Bainbridge
Editorial Assistant: Justin Billing
Program Manager: Danielle Simbajon
Project Manager: Rachel S. Reeve
Program Management Team Lead: Karen Wernholm
Project Management Team Lead: Christina Lepre
Media Producer: Jean Choe
TestGen Content Manager: Marty Wright
MathXL Content Manager: Robert Carroll
Product Marketing Manager: Tiffany Bitzel
Field Marketing Manager: Andrew Noble
Marketing Assistant: Jennifer Myers
Senior Author Support/Technology Specialist: Joe Vetere
Rights and Permissions Project Manager: Gina M. Cheselka
Procurement Specialist: Carol Melville
Associate Director of Design: Andrea Nix
Program Design Lead: Beth Paquin
Production Coordination, Text Design, Composition, and Illustrations: Integra
Software Services Pvt Ltd.
Cover Design: Studio Wink
Cover Image: RunPhoto/Getty

Acknowledgements of third-party content appear on page C-1, which constitutes an extension of this copyright page.

Library of Congress Cataloging-in-Publication Data

Agresti, Alan
    Statistics: the art and science of learning from data / Alan Agresti, University of Florida, Christine Franklin, University of Georgia, Bernhard Klingenberg, Williams College. — Fourth edition.
    pages cm
    Includes index.
    ISBN 0-321-99783-2
1. Statistics — Textbooks.   I. Franklin, Christine A.   II. Klingenberg, Bernhard.   III. Title.
QA276.12.A35 2017
519.5 — dc23
                                                                        2015016015

1 2 3 4 5 6 7 8 9 10 — RRD-W — 19 18 17 16 15

www.pearsonhighered.com

ISBN 13: 978-0-321-99783-8
ISBN 10:   0-321-99783-2

Statistics

The Art and Science of Learning from Data

Fourth Edition

Alan Agresti
University of Florida

Christine Franklin
University of Georgia

Bernhard Klingenberg
Williams College

With Contributions by
Michael Posner
Villanova University

PEARSON

Boston   Columbus   Indianapolis   New York   San Francisco   Amsterdam
Cape Town   Dubai   London   Madrid   Milan   Munich   Paris   Montréal   Toronto
Delhi   Mexico City   São Paulo   Sydney   Hong Kong   Seoul   Singapore   Taipei   Tokyo